KINDERGARTEN STEAM TALES
A Children's Book of Stories
And Adventures in 2D and 3D

S TEAMStart
S By Jeannie Ruiz

PARTNERS IN STEM PRESS

Portland, Oregon

Copyright 2014 by Jeannie Ruiz

Published by PARTNERS in STEM Press
7401 N Burlington Avenue
Portland, OR 97203

All rights reserved. No part of this publication may be reproduced, stored in a retrieval system, or transmitted in any form for any reason, recording or otherwise, without the prior written permission of PARTNERS in STEM Press.

ISBN 9781942357292

Photo permissions and authors, illustrators, designers, photographers available for download at the URL.

http://www.PARTNERSinSTEM.org

TEN80 STEAMStart DEVELOPERS ALSO OFFER *THE NATIONAL STEM LEAGUE:* EXEMPLARY SECONDARY PROGRAMS ENDORSED BY STEMWORKS

Ten80 Education and its programs are included in the STEMWorks Database by Change the Equation, a non-partisan, CEO-led initiative to connect and align efforts to improve STEM learning. Ten80 Education programs met the highest standards for excellence and exemplary STEM project based programming through rigorous examination by WestEd, an independent nonprofit research organization.

RECOMMENDED BY REAL TEACHERS

We are thoroughly enjoying our STEAM lessons. The program is excellent! My students have loved making the Rhombis and getting to build with shapes and tape. It has been a big hit! Thanks for everything!
Alicia Stenard
Mater Christi in Albany School District

"We need more projects like this that carry on for a longer number of weeks. My teachers used scope and sequence suggestions, and the program ran after school for a semester. We could have worked on this a lot longer."
Karen James
Druid Hills Elementary in Charlotte, NC

"I'm excited for our students to get their hands on real STEAM projects. There isn't much out there for Kindergarten and First Grade."
Sandy Mettler
Fort Worth Schools in Dallas, TX

KINDERGARTEN MODULE 1
WHAT IS A DESIGNER?

DESIGNERS, WILL YOU HELP ME?
- Designers collect, analyze, and synthesize data guided by specifications.

- Designers make clear and concise recommendations using drawings, verbal descriptions, and math models.

- The Industrial Designers Society of America's definition of industrial design describes design as "the professional service of creating and developing concepts and specifications that optimize the function, value and appearance of products and systems for the mutual benefit of both user and manufacturer."

Rhombi asks her friends to offer assistance in design. Students collect, analyze and make data-based decisions. They must share their concepts verbally, using drawings, and through models (scale and math).

KINDERGARTEN MODULE 1
USING THIS STORYBOOK

JUMPSTART INVESTIGATIONS WITH STORIES
Each project and STEMvestigation has an associated story. The story does not offer the recipe for success in each STEMvestigation. Rather, the literacy component is a jumpstart to discussion on the topic. Images are limited to encourage kids to offer their own vision. The goal of this book is to encourage the "bedtime story" effect. They will interact with plenty of books that walk them through with pictures. Let students tell you what they think Quad looks like or what a Saturday Farmer's Market might sell.

THESE STORIES ARE WRITTEN ABOVE GRADE LEVEL
These stories are intentionally written above the level of a kindergarten student's reading ability. Children will not build mature syntax or extensive vocabularies if they are limited to four-word sentences. Building the ability to "see" a story can only happen when the child hears the story and begins to imagine.

READ AND RE-READ
Kids don't get bored with repetition in television or books. Read the same book several days in a row. Hold a discussion each time about the images, or let kids draw their own versions. Use this book just like you'd use any compilation. You are the teacher. You know when the storybook read-aloud component is appropriate for your students.

KINDERGARTEN MODULE 1: TABLE OF CONTENTS

UNIT	READ ALOUD STEAM TALE	PAGE
CIRCLE TIME		8
SCIENCE	Three Mousy Sisters	10
TECHNOLOGY	Hickere, Dickere, Dock	12
ENGINEERING	Wheels on the Ground	14
ARTS	Circle Tree	16
MATHEMATICS	Radius and Pi	18
SQUARE ONE		20
SCIENCE	Square Foot Harvest	22
TECHNOLOGY	Fox's Forest	24
ENGINEERING	Shopping for Shapes	26
ARTS	Quilting in Circles	28
MATHEMATICS	Paper Forests	30
WE LOVE TRIANGLES		32
SCIENCE	Traveling Triangle	34
TECHNOLOGY	Butterfly, Flutter By	36
ENGINEERING	Euclid's Books	38
ARTS	City of Angles	40
MATHEMATICS	Tri, Tri Again	42
TAKE FIVE		44
SCIENCE	Hunting for a Pentagon	46
TECHNOLOGY	Tool Time	48
ENGINEERING	Creative Design	50
ARTS	Geodesic Domes	52
MATHEMATICS	The Pentagon, USA	54
BEST IN SHOW		56

KINDERGARTEN MODULE 2: **TABLE OF CONTENTS**

UNIT	READ ALOUD STEAM TALE	PAGE
RHOMBI'S PLAYHOUSE		60
SCIENCE	Wind on the Windows	62
TECHNOLOGY	Accidental Inventions	64
ENGINEERING	The Tiny House	66
ARTS	Cube World	68
MATHEMATICS	Do You Know the Quadrilaterals	70
ROOF ROOF		72
SCIENCE	Water Is as Water Does	74
TECHNOLOGY	The Case of the Disappearing Sand	76
ENGINEERING	Circle, Triangle, Square	78
ARTS	I Love Homes	80
MATHEMATICS	The Camel and the Pyramid	82
PET FINDS A HOME		84
SCIENCE	Davids Finds a Purrl / Purrl Finds a Home	86
TECHNOLOGY	Rhombi Lived in a Zoo	88
ENGINEERING	It All Adds Up	90
ARTS	M.C. Escher's World	92
MATHEMATICS	How Big Is a Guinea Pig	94
HAPPY BIRTHDAY, PET		96
SCIENCE	That's Bananas	98
TECHNOLOGY	Waxing Colorful	100
ENGINEERING	Strength in Numbers	102
ARTS	Square in the Middle	104
MATHEMATICS	Coming up a Cloud	106
RHOMBI'S CANDIES		108
GLOSSARY		110

CIRCLE TIME

Once upon a time, there was a kid name Rhombi. Rhombi liked shapes. She saw shapes all over her home and school. Rhombi's favorite shape was the circle. She saw the circle of the moon for part of each month.

Rhombi had a list of all the circles she saw. A lot of things on the list could spin. Skateboards, bicycles, and the gi-ant Ferris wheels at the coun-ty fair were all circles.

Rhombi said out loud, "All the things on my list that roll and spin are round."

Rhombi hopped on her bike and headed to Uncle Reel's bike shop. She had a bag of oranges for her uncle.

"I make sure the wheel's rotation is fast and smooth. The wheel needs to spin evenly so the bike is easy to steer," Uncle Reel said with a smile. "It is fun. You know I love things that spin."

Uncle Reel really loved things that rotated. Rhombi had an idea. She knew she would need help. She looked out from the pages of her world and asked, "Designers, will you help me cre-ate a toy that rotates?"

Three Mousy Sisters
Explore structures in science.

Point to each type of house built by my daughters...
STICKS
WATTLE AND DAUB CLAY

Three Mousy Sisters

Explore structures in science.

This is the story of three little spiny mice. One summer day when heat rolled down the rooftops to crash on dusty dry ground, Baba Mouse said to his three little darlings, "Clever daughters, it is time to make your way in the world. Build strong houses. Rain is com-ing."

The three spiny mouse sisters set off toward Baba's land on the mountainside. Oldest Sister was the first to tire of walking. She found a fl at area near the dry, dusty riverbed full of sticks. She wove the sticks together like a basket. She was lazy, so she dropped for a nap before she finished the stick house completely.

Halfway to the top of the hill, Middle Sister tired of walking. The day was hot and sunny, and she wanted to decorate her house with silk scarves from the market. She also made a stick house but had enough sense to make a wattle and daub house by filling in all the holes with mud. Before the mud daub dried, Middle Sister wandered away to pick flowers in the sun.

Little Sister walked all the way to the top of the mountain. She spent all day toiling in the hot sun. She sketched a design in the dirt and made a little mud model before beginning work on her real house. The model's roof fell, so she made a few changes in her plan. She formed clay walls, adding layer on layer until her house was finished. Its walls were strong. The redesigned roof was tight and no leak of rain would reach her tiny mouse bed of grass. The sisters settled happily into their beds that night. At five a.m. the rains came. Water dropped from the sky to pound rooftops and make the very ground shake like a drum.

Oldest Sister heard a rush of water in the dry riverbed where she'd built her house. The ground shook, but Oldest Sister escaped to scramble up the bank and run for up the mountain. Water rushed down the once-dry riverbed, taking oldest sister's house with it.

Middle Sister's house was dry. She watched the rain sluice the hill. Oldest daughter ran into the house. Rain fell hour after hour until the ground was soaked. Mud began to slide down the hill. Crash! Crash! The sisters scrambled out of the house as mud knocked down the walls. The spiny mouse sisters slipped and slid their way up the hill to the home of Little Sister.

Little Sister opened the sturdy door and hand-ed both sisters dry blankets made of rushes. She helped them find comfortable places to sleep and they waited for the sun to come again. Weeks passed as they spun wool, and read books, and ate jars of preserved food. The rain stopped as suddenly as it had begun. The mousy sisters helped each other build houses side by side at the top of the hill. That is how the mouse learned to work hard while the sun shines.

Hickere, Dickere, Dock
or The Shepherd's

Hickere, Dickere, Dock
Explore counting.

Hickere, Dickere, Dock was first published in 1777 in Ireland. Hickere, Dickere, Dock meant 8, 9, and 10 in the Westmoreland dialect of England.

Shepherds in those days used a special system of counting sheep. They could not count past 5, so their system broke numbers down into small groups of 5.

Touch each knuckle on right hand with a thumb.
That's 1-2-3-4-5.
Hand moves up a notch on the crook.

Touch each knuckle on left hand with a thumb.
That's 1-2-3-4-5.
Hand moves up a notch on the crook.

Touch each knuckle on right hand with a thumb.
That's 1-2-3-4-5.
Hand moves up a notch on the crook.

Touch each knuckle on left hand with a thumb.
That's 1-2-3-4-5.
Hand moves up a notch on the crook.

When the shepherd reaches the fourth notch, he has counted to twenty. He drops one stone in a counting bag (the clock strikes one) and drops his hand down to begin counting again.

NURSERY RHYME	HIDDEN MEANING
Hickory, Dickory, Dock	8, 9, 10
The mouse ran up the clock.	The thumb counts up the shepherd's crook.
The clock struck one.	The count reaches 20, and one stone is dropped in the bag.
Away he run.	The hand moves back down the shepherd's crook to start again.
Hickory, Dickory, Dock	8, 9, 10

Try counting like the Irish shepherds on a crook of your own.

Wheels on the Ground
Explore circles.

Wheels on the Ground
Explore circles.

Radius and Pi were best friends. Radius and Pi were very different people, however. Radius loved things that were fast. He did not stop to think about what could happen. Pi was a planner. She made lists and always made decisions after lots of think-ing. Radius and Pi made a good team.

On Saturday, the two always pulled a cart of organic fresh vegetables and fruits to the local Farmer's Market. The two grew their own food to sell. They set up a table beside all the other farmers and gardeners. Radius and Pi never missed a market.

This weekend Radius and Pi had a problem. The rain had made muddy puddles on the market path. The cart's skinny tires had sunk 2 inches in the mud. The twins were stuck. Pi sat down to think. She made notes on a pad and listed all the ways they could solve this problem. She came up with trucks and sleds.

Radius dove into his project. He liked the idea of a sled. He found a big piece of plastic and piled all the fruits and veggies on the sled. Radius yelled, "Ta Da!" He pointed to his solution and dragged the sled a few feet. It worked but was heavy.

Pi decided to try the truck idea. She found four tires from a toy in the shed. Radius helped Pi prop up the cart. They used a screwdriver and wrench to remove the cart's wheels. Radius held the cart steady as Pi connected each wheel to the cart's axle. The two lowered the cart to the ground.

Radius gently pulled the handle. He tugged a bit harder. Finally, the cart started moving. The new, wider wheels worked pretty well! Their cart sank only about 1 inch into the mud. The two moved all the produce to the wagon. The wagon rolled along the path all the way to market.

Radius and Pi sold all the food and had a great day. Best of all, they had worked as a team to solve a tricky problem. Radius and Pi always made a good team.

The Circle Tree
Explore circles.

The Circle Tree
Explore circles.

There once lived a tree in a forest up the hill. It welcomed any animal that needed a home. Its full branch-es held squirrels and birds and bees and bugs safe inside the lush green leaves. In Fall, the tree decorated the forest with its leaves of gold and red. When Winter came, the tree tucked in its branches to shelter squirrels and raccoons from snow and ice until the spring thaw.

The tree returned each Spring with a burst of leafy green branches that fed caterpillars and hid baby birds until they were ready to fly. Every animal in the forest had been helped by the tree in its 200 years. Finally, though, the tree's branches began to thin. Its bark grew brittle and broke off in great flakes. When Winter winds blew, the tree shivered. When Spring arrived, there were no leafy green hiding plac-es for tiny animals.

It was no longer home to dozens of animals, but the tree still stood tall and proud. One morning it heard singing. The tree swayed to music that played far down the hill at an end-of-the-school-year art party. Kids painted thick round paper with bright colors. They added paper flowers and rib-bons or yarn. Each child wrote the happiest word they could imagine. It was a cheerful tradition that ended every school year.

The school picnic was about to start when gust of wind snatched the school's art projects. It sent them spinning off into the forest. Laughing children and their teachers ran after the round plates. They chased the beautiful art all the way to the foot of the ancient tree. The Principal point-ed at the tree. On every branch hung a piece of art. The whole school sat under the tree to eat their picnic. Fi-nally, all the kids headed back down the hill toward waiting buses. The forest was calm again.

Singing. Cheeping. Chirping. Chitter-ing. The air around the tree was filled with happy sounds. Birds grabbed yarn for their nests. Squirrels snagged papers to line their homes.

A new tradition started. Every year, the school ended their year by hang-ing good thoughts on bright round art in the tree's branches. The tree was very happy.

Radius and Pi
Explore circles.

Radius and Pi
Explore circles.

Radius and Pi rolled along the path toward Saturday market. Suddenly, the cart stopped. Radius circled the cart looking for a problem. He leaned down to take a look at the wheel and axle. The wheel had wobbled loose and cracked.

Radius looked around for some way to fix the cart. The two needed a new wheel. Radius and Pi began looking for round objects that could fit the cart.

The two found a roller skate wheel and a skateboard wheel in a pile by the curb. Both wheels would be too small. The length from the center hole to the outer part was just too short. Their cart might roll, but it would tilt and spill the apples.

They found some balls, but none had a hole in the center. No hole meant no axle.

How would the ball stay on without an axle down the middle?

Radius tried to hold up the corner with no wheel while Pi steered. The two entrepreneurs really wanted to reach the market in time to sell their fruit. They moved too slowly without a wheel. Pi and Radius looked at the bicycles in a shop window. They looked at each other with wide grins and carefully propped the cart on a step.

Some minutes later, Radius emerged with a bicycle wheel minus the tube. It was old and rusted but not bent. The shop owner gave them the wheel. The two gave her ten apples. The wheel was just the right length from center to rim.

Radius and Pi put the wheel on their cart. They zoomed to the market. Then the market gate opened, customers were happy to see the two. Radius and Pi sold all the yummy ripe apples. The two were proud that they'd solved an engineering problem all on their own.

UNIT 2
SQUARE ONE

SQUARE ONE
Explore squares.

Once upon a time, there was Rhombi. Rhombi loved shapes. Rhombi had been curious about circles. She had taken pictures of circles and made art with circles. She had even made a gift with circles.

She was very proud of the spinning gift she'd made for her uncle. Her designer friends had made models that helped her create the round rotating toy.

Rhombi was playing with a cardboard barn and horses. In her game, the farm's goat stomped on the barn roof. Her hands made the goat jump up and down.

Suddenly, the barn's roof cracked right down the middle. Her toy barn's roof hit the floor and broke into pieces. Rhombi was upset. She liked the barn a lot and wondered how she could fix the roof.

Rhombi knew that the roof had four edges and four corners. She was pretty sure all the edges had been the same length. That made the roof a square. She did not know the roof's material. Rhombi had an idea.

Square Foot Harvest
Explore Garden Harvests in Science

Square Foot Harvest
Explore Garden Harvests in Science

Leaves crunched underfoot as Rhombi headed toward the garden. Rhombi and her friends had planted vegetable seeds in the Community Garden in August. Over the past two months, the seeds pushed out of their hard seed covers. In September, the seedlings climbed through dirt toward the sun.

Rhombi had spent many hours making sure that each seedling had sun and room to grow. Then she saw leaves, Rhombi knew that roots had reached down into the soil. Finally, after almost days, the radishes were ready to eat.

Carrots and radishes and lettuce waited for Rhombi. She pulled 2 carrots and put them into her basket. She layered 2 radishes. Finally, Rhombi added four heads of lettuce.

She picked up her heavy basket and headed home through the crunchy leaves.

She washed and chopped carrots and radishes. She tore lettuce leaves and added them to a big bowl. When her friends sat down to dinner, Rhombi would proudly serve her delicious autumn har-vest salad!

Fox's Forest
Explore environments in art.

Fox's Forest
Explore environments in art.

Fox trotted happily on the warm summer path. Sun beamed down through leaves to make moving patterns on the worn dirt track. Fox's head swung from side to side looking for yummy treats. He also scanned the woods for anything that might want to eat him!

Bushy tail wagging, Fox stopped to nose some flowers. Soon these green leaves would part to reveal little red berries. Delicious and sweet, Fox loved these berries in the straw. Fox caught another smell in the straw. It was the ripe odor of rotting log. Logs were good. They often held tasty, crunchy bugs.

Fox lifted his little pointed face to the wind. Here was the smelly log He faced west toward the setting sun. Sniff, sniff. Fox smelled logs there. He faced east toward the far edge of the woods. Log smell rose from that direction. Fox turned north to sniff. There. The stinky log was that way!

Nose to the ground, Fox left the path and headed into the green woods. Shadows played across Fox's rust colored body. Sniff, sniff. He followed the air molecules that blew in the breeze from that stinky log. The molecules were pulled in through Fox's nose and mouth. His brain made sense of the smells, and Fox turned a little to the east. Eventually, behind a tall rock, Fox found a log so rotten it made his mouth water. Using tiny paws and claws, Fox pulled at the wet bark. Beetles and meal worms tried to scatter. Fox caught them and crunched his way through an afternoon snack. Delicious!

With his pointy little face covered in soil and bits of rotten tree, Fox turned back toward the path. South and a little East brought Fox right back to strawberry leaves beside the warm summer path. It wasn't quite so warm now that the Earth was turning away from the Sun. Shadows crept closer as Fox found his way back to the den. With a full belly, Fox curled up and fell asleep.

Shopping for Shapes
Explore shapes in engineering.

Shopping for Shapes
Explore shapes in engineering.

Poly and her friend Trap were shopping for groceries. The two were meeting up with Rhombi that afternoon for chess club. It was Poly's turn to bring snacks. She had ten dollars to spend.

Poly pushed a cart up and down the aisles looking for something her friends might like to eat. She dropped a pack of gum in the cart, but it fell through the squares that covered the cart. She placed the gum on the rectangular face of a box of cookies.

Poly wanted to find cheese, crackers, fruit, and cookies. So far, she'd found cookies. They were the round kind with chocolate chips. Trap held up a bag of $3.00 star fruits. They looked like spheres or balls now. When cut, the star fruits would look like little green star shapes on a plate. He also had fat purple grapes that cost $1.00 per bag.

The list included cheese, crackers, fruit and cookies. So far, Poly and Trap had cookies and fruit. The two friends headed toward the dairy section to look for cheese- Wow, there were a lot of cheese shapes. Which one should they choose? There were cylinders of string cheese. Poly pointed to brick shaped rectangular prisms of cheddar. Trap spied a bag of already cut cubes that would be easy to pile on a tray. The $2.00 cheese cubes were added to the cart.

The last item on their snack list was crackers. The cracker aisle was filled with boxes of all shapes and sizes. Oval buttery crackers sat next to 2 inch square saltines and little ½ inch round soup crackers. There were whole wheat rectangles woven like blankets and cheesy 1 inch squares. Some were even shaped like goldfish. Trap picked the $2.50 woven rectangles. They tasted good with cheese.

Their cart was filled with delicious shapes. Star fruits, grapes like spheres, cubes of cheese, and square crackers filled the cart. The total cost was $8.50. They had one dollar and 50 cents left. Poly and Trap took their bags and headed to the chess club. Snack time shaped up to be a yummy break.

Quilting in Circles
Explore circles.

Quilting in Circles
Explore circles.

Mole and Mouse are sewing a quilt. They will need to stack three layers, one on top of another. The friends start with a bottom layer using squares of old t-shirt fabric. The center layer is fluffy cotton from pillows. The top layer is bright fabric with circles.

Each friend has a thimble to keep fingers safe from sharp needles. They use a pattern to cut many circles in four different sizes. Mole cuts circles that measure 1 inch from the center, 2 inches from the center, 3 inches from the cen-ter, and 4 inches from the cente.

Mole draws a pattern of circles that look like a rainbow of dots on the fabric. Some of the circles touch edge to edge. Other circles sit one inside another. Quilters call this pattern an echo. Each circle is sewn by hand. Each circle is added one by one.

The planet is rotating away from the Sun when Mole and Mouse finish the project. The Sun sink round full moon in the sky. The twins put down their needles. They shake out the colorful quilt. Circles overlap and dance on one side. Bright t-shirt squares fill the other side.

On the day of their library fund-raiser, Mole and Mouse sell raffle tickets. Friends and family donate a dollar and get a raffle ticket in return. At the end of the day, someone will win the quilt made by Mole and Mouse.

Mole and Mouse sell 100 raffle tickets and make $100 for the library. At the end of the day, Mole pulls one of the tickets out of the bowl. Mole starts to giggle. Finally, she calls out the winner's name. It's Mouse! Mouse has won her own quilt.

Paper Forests
Explore origami in mathematics.

Paper Forests
Explore origami in mathematics.

Rhombi's Grandpa folded beautiful paper art. When Rhombi was little, Grandpa could make animals and flowers and boats and castles appear from his hands using nothing but squares of thin paper. Grandpa taught her to fold paper boxes to hold paper jewels. He taught Rhombi to make beautiful princesses with swords that rescued dragons from wicked wizards. Rhombi's favorite project was a mobile hung with 40 folded animals. Grandpa called it his forest in the sky.

Grandpa no longer folded paper, though. His hands were old and gnarled. The fingers couldn't hold delicate paper anymore. Now, instead of the country, Grandpa was living in a city home with lots of other grandmas and grandpas. They all played checkers and had holiday parties. Grandpa seemed happy, but sometimes he talked to Rhombi about the old days. They laughed about a funny fast squirrel that sneaked in through the window and raced around the bedroom before burst-ing out the front door.

Grandpa remembered pushing Rhombi on the tire swing. He talked about animals that visited from nearby woods. Grandpa especially missed the red fox that used to peer in through the bedroom windows on cold snowy nights. Sometimes they found it sleeping in his shed. Grandpa's new home was in the city. No animals peeked in city windows.

Rhombi planned to visit Grandpa for his birthday on Saturday. Sitting at her desk, Rhombi thought about a gift. She knew he missed his forest and fields, but Rhombi could not bring any live animals into Grandpa's new home. Looking at all the origami papers on her table, Rhombi decided that she could bring some animals to Grandpa after all. She picked up a stack of papers and started to fold.

Saturday was the big day. Rhombi eased her gift through the front door of Grandpa's building. Other grandmas and grandpas and aunties and uncles smiled as she made her way to the dining room. All Grandpa's friends were there with cake and balloons. Everyone oohed and ahhed as Rhombi entered. Grandpa turned to look, and his hands flew to his mouth.

He smiled. He grinned. He laughed out loud! Hanging from Rhombi's hands was a mobile full of forest animals. His squirrel was there in its tree. There were blue birds and rabbits. But, best of all, hanging bright and sassy at the top of Grandpa's new forest in the sky was a beautiful bright red fox. Happy birthday, Grandpa!

UNIT 3
WE LOVE TRIANGLES

WE LOVE TRIANGLES

Explore triangles.

Once upon a time there was Rhombi. Rhombi loved shapes. Circles were cool (there were circles in the word "cool"). Her innovative helpers had designed a tool to find squares. That had been a fun project.

Rhombi sat in the park. She looked around the playground and played her own game of "I SPY" with shapes.

"I spy with my little eye a square tile."

"I spy with my little eye 80 circles."

80 circles were linked to make a chain that held the swing. She had counted them one day at recess. The swing seats were squares. She knew because each seat had 4 equal sides and 4 corners shaped like the capital letter L.

Rhombi saw triangles in the swing. The triangles held up the swing set and kept it from tipping. Rhombi decided that the triangle must be pretty strong. She watched some kids swinging really hard, but the triangles did not fall. Rhombi saw triangles in the slides and park bench, also.

After recess, Rhombi's class had center time. Rhombi went to the hamster cages to clean the cedar. She would refill the pets' water, too.

The hamsters sat on their nests. The habitats were clear plastic so kids could see the hamsters. The poor rodents looked bored inside the 3D homes.

Kids used recess to recharge their brains. Rhombi wished she could give the pets recess. Rhombi smiled and said, "Well, if I can't take you to recess. I can bring recess to you!" Rhombi looked out the window. She saw swings on the playground. Then Rhombi saw the slides.

Rhombi knew she would need some help. She looked out from the pages of her world and asked, "Designers, will you help me create a slide with a sturdy triangle base for our hamsters? They weigh about as much as a beanbag."

Traveling Triangle
Explore triangles in engineering.

Traveling Triangle
Explore triangles in engineering.

Triangle, Triangle,
where have you been?
I've soared in the sky
blown East and West by the wind.

Triangle, Triangle,
what have you seen?
I've watched the city
from the top of a Ferris wheel.

Triangle, Triangle,
what have you done?
I've held the wood steady for
builders in the hot summer sun.

Triangle, Triangle,
where have you gone?
I've biked along city streets
in the cool quiet dawn.

Triangle, Triangle,
what should we play? Let's
climb on the jungle gym
at the end of the day.

Butterfly, Flutter By
Explore symmetry.

Butterfly, Flutter By
Explore symmetry.

She eats candy.
Candy eats she.
He loves chocolate.
Chocolate loves he.

They share sunshine.
Sunshine share they.
They watch butterflies.
Butterflies watch they.

Butterflies flutter.
Flutter butterflies.
Butterflies are shy.
Shy are butterflies.

Wings move swiftly.
Swiftly move wings.
Goodbye butterflies.
Butterflies goodbye.

Euclid's Books
Explore the history of geometry.

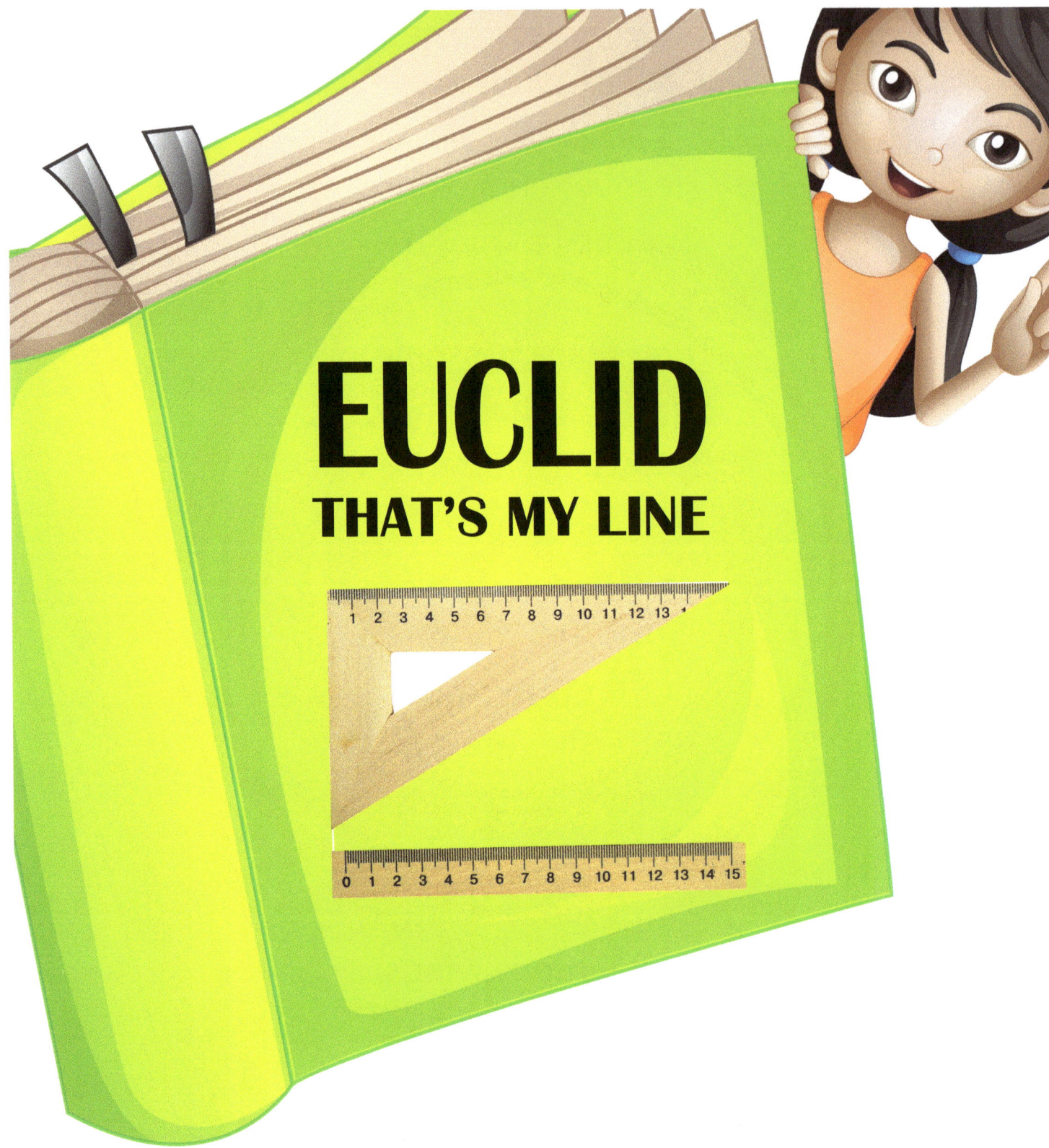

Euclid's Books
Explore the history of geometry.

Euclid (YOU-klid) was one of the earliest scientists. He wanted to use reason and logic to prove the truth of things. Copies of Euclid's books can be found in libraries today. The books begin with definitions of a point, a line, and shapes. Euclid used geometry to prove that all right angles are equal. That means all the angles that make up corners of paper or corners of squares are shaped the same.

Euclid taught even more interesting math. He wrote about how triangles and circles work. He also showed that flat shapes could be put together to make not-flat shapes. For example, six flat squares can make a cube.

Euclid is famous because his writing was easy to understand. For two thousand years, Euclid's books were used as the math book around the world. Some mathematicians still working today started by studying geometry from Euclid's books.

City of Angles
Explore maps in technology.

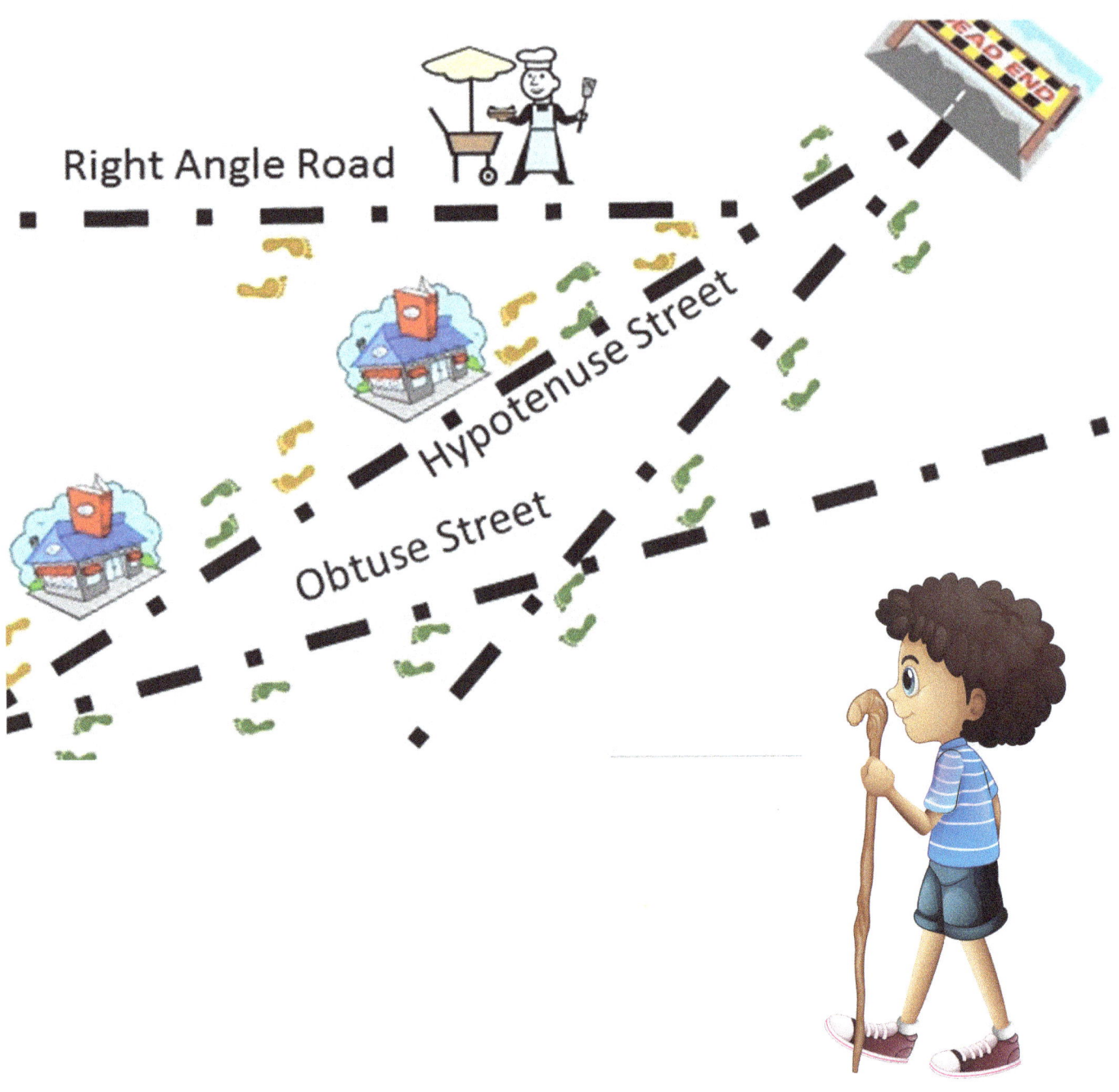

City of Angles
Explore maps in technology.

Every street in Three Angles was named after the shape of its corner. That made walking through town easy if you knew your shapes. Quad wanted to find Polygon Museum. He was new to the city, but he was pretty sure he could get around. Quad didn't bother taking a map.

Quad walked North along Line Street to the town square. He turned right at the square and walked East along Right Angle Road. Quad bought a cool dog collar at the Pet Palace. At the farthest end of Right Angle Road, Quad took a sharp right turn and headed South along Hypotenuse Street. He found two great bookstores along the way. Finally, Quad ended up right back where he'd started! He had not found the museum. He decided to try again on Friday.

The next morning, Quad looked for the Museum again. This time, he started on Hypotenuse Street. He headed North on Hypotenuse until the street came to a dead-end. Quad decided to turn right onto Acute Avenue. He was sure to find the museum at the end of Acute Avenue. Quad reached the corner and took a slight right onto Obtuse Place. He walked and walked until he saw the sign for his hotel at the end of Obtuse Place just before Hypotenuse.

Good grief! The sun was setting. He'd have to find the museum on Saturday. This time, he'd save time and use a map.

On Saturday morning, Quad stood outside his hotel. The young tourist looked at his map. He touched the name of the Polygon Museum on the map. Slowly, Quad looked left. The museum was beside his hotel. Had Quad turned left out of the hotel, he would have seen the Museum right away. Next time, Quad would definitely use a map.

Tri, Tri Again
Explore triangles in mathematics.

Tri, Tri Again
Explore triangles in mathematics.

Poly really needed to fix her bird feeder's roof. The feeder hung from a branch outside her window. All kinds of feathered friends relied on the feeder for their winter meals. Sadly, there were a bunch of furry friends that also wanted to eat at the hanging seed source. Long fluffy tails knocked the feeder to the ground over and over again. Poly was afraid that the feeder would break in one of its squirrelly falls.

Poly had tried lots of different ways to trick the squirrels. She wanted to keep them away from the feeder, but the mammals wanted all the free seeds. Squirrels had nuts of their own, but the bird feeder was such an easy food source. Poly had designed a round tipsy cover that should have made the squirrels wobble and fall to the ground. They learned how to climb around it in less than a week. Poly had turned a mesh scarf into a wrap through which squirrels couldn't get their hands. They'd just untied and dropped the rectangular scarf to the ground. Poly had even covered the bird feeder hanger with drops of slippery oil. The greedy beasts just slid down it like a firefighter's pole. They even seemed to have fun!

Poly had thought she was out of ideas. Round plates hadn't worked. Rectangular scarves hadn't worked. Spheres of oil hadn't worked. That afternoon, on her class trip to the city museum, Poly had seen a tall pointed roof. The roof covered a bell, and the guide told her class that the roof kept pigeons from landing. Bingo! If it kept pigeons away, the roof should keep squirrels away.

Poly cut out 3 triangles of wood. She covered them with waterproof material used for raincoats and connected all three triangles at their long edges. She formed a tall pointed object and attached it to the bird feeder. Her brand new triangular roof was nearly 3 feet tall and very steep. She hung the modified feeder and sat down to wait.

Birds flew in and out to get their seed meals. Finally, two jolly little squirrels came to sit on the branch above the feeder. They tilted little heads to one side and the other side looking confused. Finally, one brave little friend leaped to the roof. She slid right past the seed and landed on the ground. That was one grumpy squirrel. Her rodent friends each tried the feeder, but all failed.

Poly had learned her lesson. If at first you don't succeed with circles and rectangles, tri, tri again.

UNIT 4
TAKE FIVE

TAKE FIVE
Explore pentagons.

Rhombi, Quad and Poly put on hiking boots. They fill bottles of water. All three friends wear long pants and long sleeved shirts. Quad hands each friend some sunscreen, and they all put on a hat.

The friends will look for deer, snakes, raccoons, birds, and turtles. The kids open a map of Pentagon Park with 5 edges and 5 vertices. It's a pentagon.

"Hey, the Nature Center trails are laid out in a pentagon," Poly shouts. "That's like the shape on my soccer ball."

Quad laughs but shushes Poly. "We will never see any animals if you shout."

The kids take a trail leading to the deer meadow. With any luck -and quiet voices- they will see some animals in the woods. They hike for half an hour.

"Look at the deer," whispers Poly.

Peeking from the edge of the woods are two deer. Many deer have used this place to cross the path. They are perfectly still and silent. Suddenly, there is a commotion. Mountain bikers speed down the trail toward the deer. Poly shouts at the deer to move faster. She shouts at the bikers to stop!

The bikers have permits and are being safe, but an accident could happen any second! The mountain bikes skid to a stop. Gravel and dirt fly into the air. The last deer bounds away into the trees.

"There should be a sign to let us know this is a place where deer cross the path. If this is a place the deer walk a lot," says one biker. She is breathing hard.

Rhombi decides to ask for help on this project. She looks out from the pages of her world and asks, "Designers, will you create deer crossing signs like the school crossing signs outside your school?"

We're Hunting for a Pentagon
Explore pentagons in science.

We're Hunting for a Pentagon
Explore pentagons in science.

We're going on a 5 hunt. (We're going on a 5 hunt.)
We're looking for a pentagon. (We're looking for a pentagon.)
We're gonna catch a big one. (We're gonna catch a big one.)
I'm not afraid. (I'm not afraid.)
Are you? (No way!)

Here comes the crossing sign.
Better wait for the crossing guard.
Walk briskly and pump your arms.

REFRAIN

We're passing a tall house. It sure is wide and has 5 edges. Let's walk around it.

REFRAIN

We're walking through the soccer field. Uh-oh. Here comes the soccer ball. We better duck under it…

REFRAIN

We're running through the garden now. It sure is grassy. I see the morning glories and okra. Jump over the rows. Oooohhh. We're all muddy. Shake it off….

REFRAIN

I see a big building. It has 5 sides. It's The Pentagon.
Oh no, what time is it?

Recess starts in 5 minutes. We're going to miss it.
Better get home fast!
Oh no, mud! Shake it off.
Jump the okra and morning glories.
Yikes, duck! Here comes the soccer ball.
Walk quickly around the tall house.

STOP! The crossing guard says to wait. Okay, we can cross
now. Walk briskly and pump those arms. Hurry! Hurry!
Ahhhhh, we made it!

Tool Time
Explore tools in technology.

Tool Time
Explore tools in technology.

Hilda and Poly had a job to do. Grandma wanted the garage organized. She liked everything in its place. The two friends had taken out the trash. They had bagged cardboard, newspaper, glass and cans for recycling. They had even swept and dusted.

The tool bench was still a big mess. Tools were scattered on the table. Tools were dumped in boxes under the bench. Tools were hung in all the wrong pegs on the wall.

Hilda and Poly started by sorting the tools used to move items. They hung hammers and mal-lets on the peg board over jars of nails in lots of sizes. Star-shaped Phillips head and flathead screwdrivers used to tighten or loosen things went into bins by size. #1, #2, and #3 were matched to their same sized screws.

Rhombi put grabbing, twisting, and pulling tools on the pegboard. Pliers were hung from smallest to largest. Wrenches were hung over jars of nuts and bolts.

Rhombi and Quad did not touch the cutting tools. Saws, hacksaws, pincers and wire cutters were left for Grandma to handle later.

Carpenter's levels, rulers, angles, squares, and other measuring tools were hung neatly to the right of the pegboard wall behind the bench. A caliper used to measure very tiny items was placed in its box. Some tools were delicate.

When Hilda and Poly finished the job, they proudly called Grandma to see her shiny garage. The excited old woman clapped her hands and hugged both girls. When Grandma offered them money for doing the job, Hilda refused to take the money.

"No, Grandma," said Hilda. "I'm happy to help!"

Creative Design
Explore pentagons in engineering.

How Many Pentagons Can You Count In This Set Of Pictures?

Creative Design
Explore pentagons in engineering.

Pentagons have 5 edges, 5 edges, 5 edges. Pentagons have 5 edges,
regular or not.

Regular pentagons have 5 edges all the same, all the same, all the same. Regular pentagons have 5 edges all the same length.

Irregular pentagons have 5 edges not the same, not the same, not the same. Irregular pentagons have 5 edges
NOT the same size.

ALL pentagons have 5 angles, 5 angles, 5 angles. All pentagons have 5 angles
regular or not.

The angles of all pentagons add up, add up, add up. The angles of all pentagons add up to 540 degrees.

5 Edges
5 Angles
500 + 40 Degrees
PENTAGONS are fun!

Geodesic Domes

Geodesic Domes

Young man, you amaze me. --*Albert Einstein, physicist (1938)*

Buckminster Fuller said, "When I draw a circle, I immediately want to step out of it." Buckminster Fuller was an architect and an en-gineer born in 1895. He was also a designer and an inventor. Full-er wrote 28 books on subjects in math and science. He had 47 special degrees from different uni-versities.

Fuller was someone that cared about making the world better. He was also a humanitarian, someone that wants to improve life for other people. Each year the Buckminster Fuller Institute offers a $100,000 prize to "scientists, students, designers, architects, activists, entrepreneurs, artists and planners from all over the world using in-novative solutions to solve some of humanity's most pressing prob-lems." From a floating doctor's of-fice to a new plan for food production, Fuller's prize helps people with great ideas afford to put their plans in action. This inventor was interested in using the planet's resources more effi-ciently. He invented a geodesic dome made of polygon shapes. The dome uses fewer materials and can be made bigger or smaller to meet the needs of many different types of uses. Gardens, homes, and museums are shaped like geodesic domes.

When Buckminster Fuller first imag-ined the special dome, he thought of it as humans working like a crew to survive on our planet, "spaceship earth."

Epcot Center in Florida created a domed building inspired by Fuller's idea. Domes like these are made of triangles. As more triangles are added, the shape gets stronger. No other 3-dimensional shape can be so big without extra supports inside the building. The larger it is, the stronger the dome becomes. These domes are so simple that an entire house can be built in a single day. They are so strong that geodesic domes can even survive hurricane force winds!

Can you see the triangles in the playground dome? Can you see dia-monds in the playground dome? Can you see hexagons in the playground dome?

The Pentagon, USA

HOLDS A LOT OF CARS

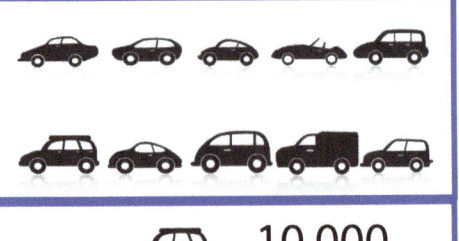

🚗 = 10,000

Aerial Image by David B. Gleason from Chicago, IL

on the PENTAGON

1. formal dining room
2. cafeterias
3. days to design the building
4. thousand 2 hundred clocks
5. edges and vertices in a pentagon
6. snack bars
7. minute walk corner to corner
8. decades since building opened
9. hundred 21 feet per outer edge
10. thousand visitors a year

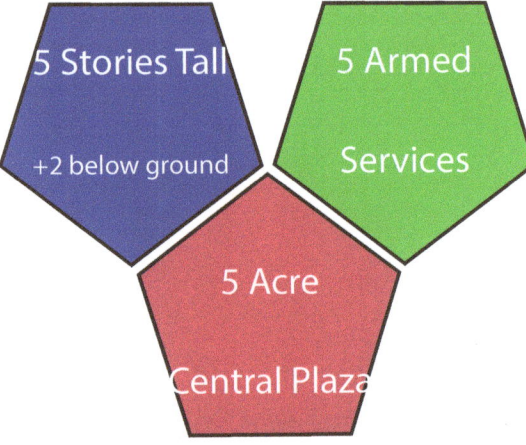

- 5 Stories Tall +2 below ground
- 5 Armed Services
- 5 Acre Central Plaza

RESTROOM

Executive Order 8802 let The Pentagon open all its 284 bathrooms to all employees, black or white. For a time, The Pentagon was the only non-segregated building in Virginia.

NUMBER OF MONTHS FROM GROUND BREAKING TO OPENING CEREMONY											
	JAN	FEB	MAR	APR	MAY	JUNE	JULY	AUG	SEPT	NOV	DEC
1941									■	■	■
1942	■	■	■	■	■	■	■	■	■	■	■
1943	■										

The Pentagon, USA

Infographic is short for information graphic. An infographic uses pictures and words to make lots of information and numbers easy to grasp. Look at this infographic showing information about The Pentagon. You could read the paragraphs that follow, or you can use an infographic to show the same information.

The Pentagon was designed in just 3 days and built far faster than most government buildings. It was also built to hold a lot of cars. Parking was planned for 10,000 cars. 16 months passed from the first day of construction until employees parked all those cars in the lots. The construction trucks, earth movers, and builders showed up for work on September 11, 1941. The Pentagon was completed in January of 1943. The building opened almost 8 decades ago.

The building serves a lot of needs. Very special visitors are invited to par-ties in the formal dining room. Most armed services workers eat in one of the 2 cafeterias. 5 different armed ser-vices use the building. 10,000 or more visitors use any of the 6 snack bars.

Finding a bathroom in The Penta-gon is easy. When the building was planned in 1941, Virginia state law said that black people and white people had to use different. The pentagon was built with 284 bathrooms. By order of the President, none of those bathrooms were ever really set aside for any special group of people. Employees and visitors to The Pentagon all used any of the public bathrooms. For a time, The Pentagon was the only non-segregated building in the state of Virginia.

Getting from place to place in the huge building takes time. Each outer edge measures 921 feet. There are 5 stories full of offices and a 5 acre central area called "ground zero." The total square footage of the The Pentagon is more than 5 million square feet. Your school probably measures less than 50,000 square feet. Walking from one corner to another corner of The Pentagon takes nearly seven minutes. How many seconds pass if you walk from one corner of your building to another corner of your building?

Which format seems easiest to use? Which format gives information quickest? Which format offers more of a story? Do you like the paragraphs or graphic style best?

BEST IN SHOW
Extension for Module 1

Once upon a time, there was Rhombi. She had very good friends that helped her design and create. Her friends had made spinning toys. They had designed innovative ways to fix the square roof on her toy barn. Friends had built a slide for the hamster from triangles. They had even designed signs to keep deer and mountain bikers safe in the woods.

Rhombi and her friends stood outside the local animal shelter. The shelter took care of pets whose owners could no longer care for them. Dogs, cats, rabbits, guinea pigs, birds, hamsters, snakes, and even a few lizards were all waiting for new families.

Rhombi, Quad, Pi, Radius, Ginny, and a few other kids were planning to help the shelter animals. They hoped to make life better for the animals until every pet found its "forever home."

"We need to decide on a project," said Rhombi.

Quad thought raising money with a bake sale was a great idea. Pi and Radius remembered adopting their dog from the shelter. The dog had been clean but scruffy. They suggested holding a pet wash. Everyone grinned at the idea of giving some angry cat a bath.

"Maybe a pet wash is not the best idea," giggled Pi.

Pi's phone beeped. She stepped away to answer the phone. When she ended the call, Pi was jumping up and down with a huge grin on her face. Pi was waving her hands around and talking quickly. She shouted, "My aunt just called. She just bought a super cool new rabbit hutch, and it gave me an idea. What about making new kennels and cages for the animals?"

Everyone agreed that this was a great idea. They would need lots of help planning and building. Rhombi and her friends looked beyond the pages of their world to ask, "Designers, will you work in teams to help us plan innovative new habitats for the animals in the shelter?"

UNIT 1
RHOMBI'S PLAYHOUSE

RHOMBI'S PLAYHOUSE
EXPLORE SHAPES

Once upon a time, there was Rhombi. Rhombi loved shapes and found them everywhere. She especially loved the squares that made up each face of a cube.

Rhombi had a new tool set, and she wanted to build a playhouse. Rhombi chose a shape for her house that looked like blocks and sugar cubes.

Rhombi knew that she should build a sturdy house to withstand the fall winds, but the she also wanted to go apple picking. Rhombi hurried to finish the playhouse so she could run and play.

That night, the weather changed. The air grew colder and howled wildly around the corners of Rhombi's home. She walked to the edge of the field to watch the rainstorm.

Returning home, she jumped in surprise as one sharp gust of wind pushed at her house. Another gust of wind pulled at the house. The walls shuddered. The house shivered. So did Rhombi.

She stared at all the crispy leaves as they tumbled from the trees. Then, Rhombi stared in amazement as the walls of her house also tum-bled to the ground.

Looking at the walls piled on the cold ground, Rhombi decided she needed some help. She looked out beyond the pages of her world and asked, "Designers, can you help me build a new and stronger house?"

Wind on the Windows
Explore weather in science.

WORDS FOR WIND

CHINOOK
westerly wind off the eastern side of the Rocky Mountains

SANTA ANA
easterly towards Southern California

SCIROCCO
southerly from North Africa to southern Europe

MISTRAL
northwesterly from central France to Mediterranean

MARIN
southeasterly from Mediterranean to France

BORA
northeasterly from eastern Europe to Italy

GREGALE
northeasterly from Greece

ETESIAN
northwesterly from Greece

LIBECCIO
southwesterly towards Italy0

Wind on the Windows
Explore weather in science.

My dog is scared. She is shaking and whining under the bed. I sit on the rug and rub her ears. It does not seem to help. Daisy is a big floppy German Shepherd, and she is not scared of anything. Well, she is scared of one thing. Daisy hates the sound that wind makes as it hits the corner of the house.

I try to tell her it is just the air moving. I speak calmly, and I stay very quiet and still. All these things help Daisy feel safe. I describe the sounds I hear. There are four tall pine trees outside the window. The wind through their needles sounds like a quartet of violins. Air pushes past the pine needles like a bow on violin strings. A willow tree by the creek trails its long branches into the water. Air lifts the branch-es and whistles through leaves like a dozen flutes singing.

Daisy is shaking a little less. I think she likes my story. Her head rests on my leg, and big brown eyes look up at me through tan lash-es. Her tail thumps a bit. I keep talking. Sound starts as vibration.

I thump Daisy's name tag to make it jingle. Sound hits your ears like a pendulum swinging. Little tiny hairs in your ears start to vibrate. All those branches and leaves shaking outside are vibrating the air, see? Daisy's eyes roll to the window, but she does not lift her head. Faster vibration means high pitched sound. Listen.

We close our eyes and listen to the wind. All that moving air pounds the window panes, making each piece of glass rattle like a drum. Some of the wind slips through the cracks around the window, and we hear deep mellow sounds. "Those are drums and cellos in our symphony," I tell Daisy. She is not impressed.

Windy flutes, violins, cellos, and drums play. With the moon high above and the faded rug below us, Daisy and I drift off to sleep. Some wild and blustery version of Bach's Cello Suite No.1-Prelude flies through the night. Daisy calms down as we listen to the symphony of wind.

Accidental Inventions
Explore new ideas in technology.

1.0  POPSICLES

11 year old Frank Epperson wanted to make soda pop at home. In 1905, the popular drink could only be bought at stores and restaurants. Frank used his porch as a testing lab. He mixed powder and water to find the right taste and tex-ture. Frank left the ingredients in a cup where he'd been working.

Temperatures dropped overnight. Frank's drink froze with the stir stick stuck in its center.

Soda Pop + Icicle = popcicle

2.0  WAFFLE CONES

The 1904 World's Fair is the birthplace of today's popular ice cream cone. Ice cream in dishes had been around for years. A dessert seller at the Fair was doing so much business that it quickly ran out of bowls and plates. A waffle maker at the next booth was not doing much business at all. The two business owners worked together to roll up the Persian waffles and fill them with ice cream.

Ice Cream + Waffles = Ice Cream Cone

3.0  CHOCOLATE CHIP COOKIES

The Toll House Inn's owner was named Ruth Wakefield. Ruth wanted to make some of her delicious chocolate cookies for people staying at the inn. Ruth realized that she was out of baker's chocolate. The inn keeper broke the regular sweet chocolate bar candy into little chips for the cookie dough. She expected the bits to melt into chocolate cookies. They did not. Instead, the chips stayed whole.

Dough + Chips = Chocolate Chip Cookies

4.0  STICKY NOTES

In 1968, Inventors Spencer Silver and Art Fry were researchers at 3M Laboratories. Spencer created a "low tack" sticky substance. It was unique in that it could be removed without hurting a surface. Unfortunately, no one had a use for the stuff. It went into storage. Many years later, Art Fry needed a way to keep papers in his choir's song book. He suggested a use for Spencer's sticky invention!

Low Tack Sticky + Paper = "Post-Its©"

5.0  SLINKY SPRING TOYS

Naval Engineer, Richard Jones, dropped one of the tension springs for a battleship project on the ground. It bounced from place to place all over the room. He spent about 2 years finding just the right materials and coil to turn his spring into a toy. Betty, his wife, named the new toy "Slinky." Slinky is a Swedish word that means sleek or sinuous. In 1945, the first batch of 400 toys sold out in 90 minutes.

Coil Spring + Hard Work = Slinky ©Toys

6.0  ?

What items do you use that might be combined to create a new invention?

Could you find a new use for something that you already own?

What would you invent to solve a problem or make the world a better place?

The Tiny House
Explore structures in engineering.

The Tiny House
Explore structures in engineering.

My Grandpa and I are building a tiny house. Grandpa says his tiny house will use less electricity and water than the house he and Grandma used to share. I am helping. Most of Grandpa's tiny house will be built far away and delivered to him in big flat boxes. We will put all the different pieces together to make a whole house.

There is a lot to do before the boxes get here. First we prepare the site. The land must be "graded." All the rocks and roots are taken away. The land is made as flat as possible. We drive a little road grader back and forth, then use a laser to level the ground. Next comes the foundation.

Grandpa forms and pour concrete cubes called pier pads. We have to wait a couple of days while the concrete sets. Two 4ft x 8ft beams rest side-by-side on the concrete. This holds the floor underneath the walls and house. I use a wrench to thread nuts onto bolts. Grandpa follows behind to tighten each one. He lays the drainage pipe and wiring through floor tunnels before adding the last of the sub-floor.

Grandpa will use a special composting toilet. All waste runs into a system of barrels under the tiny house to become fertilizer for his garden. The water will come from city plumbing, but grey water (waste water from sinks) will drain into the garden also. Grandpa uses a generator to power tools for framing and roofing. Grandpa's walls were pre-made. Aunt Elizabeth comes over to prop up the walls and attach frames. We "plumb" the walls to be sure they stand up straight from the floor. I hold a "level" tool so that its bubbles are perfectly centered. We make sure windows and door are level and square. We wrap it all in sealing tape to keep out water and mold.

Next, we'll add the roof. Grandpa uses a band saw to cut the rafters that rise from walls to roof. The rafters are spaced 4 feet apart. Insulation is added to walls and roof, then layers of roofing material top off the tiny house. Grandpa and I paint the tiny house and add shutters. We hang a sign that says, "Welcome to Grandpa's Tiny House in the Woods."

Cube World
Explore cubism in art.

Cube World
Explore cubism in art.

In the early twentieth century, artists began to look at the world as geometry in motion. Cubism began in France in 1907. Picasso and Braque started painting the world made up of cubes, spheres, cylinders, polygons, cones, and other geometric shapes. Cubists wanted to show all the sides of an object in the same picture. The paintings looked a lot like the artist had cut the image and glued its pieces back together.

Cubist paintings show objects from more than one angle at once. Pablo Picasso and Georges Braque were first to add bright colors to cubist art. They believed that painters should not just repaint the world exactly as it exists. Instead, they wanted to show every part of the whole subject.

How does the world change when you re-imagine everything as two-dimensional and three-dimensional shapes?

Do You Know the Quadrilaterals
Explore quadrilaterals in mathematics.

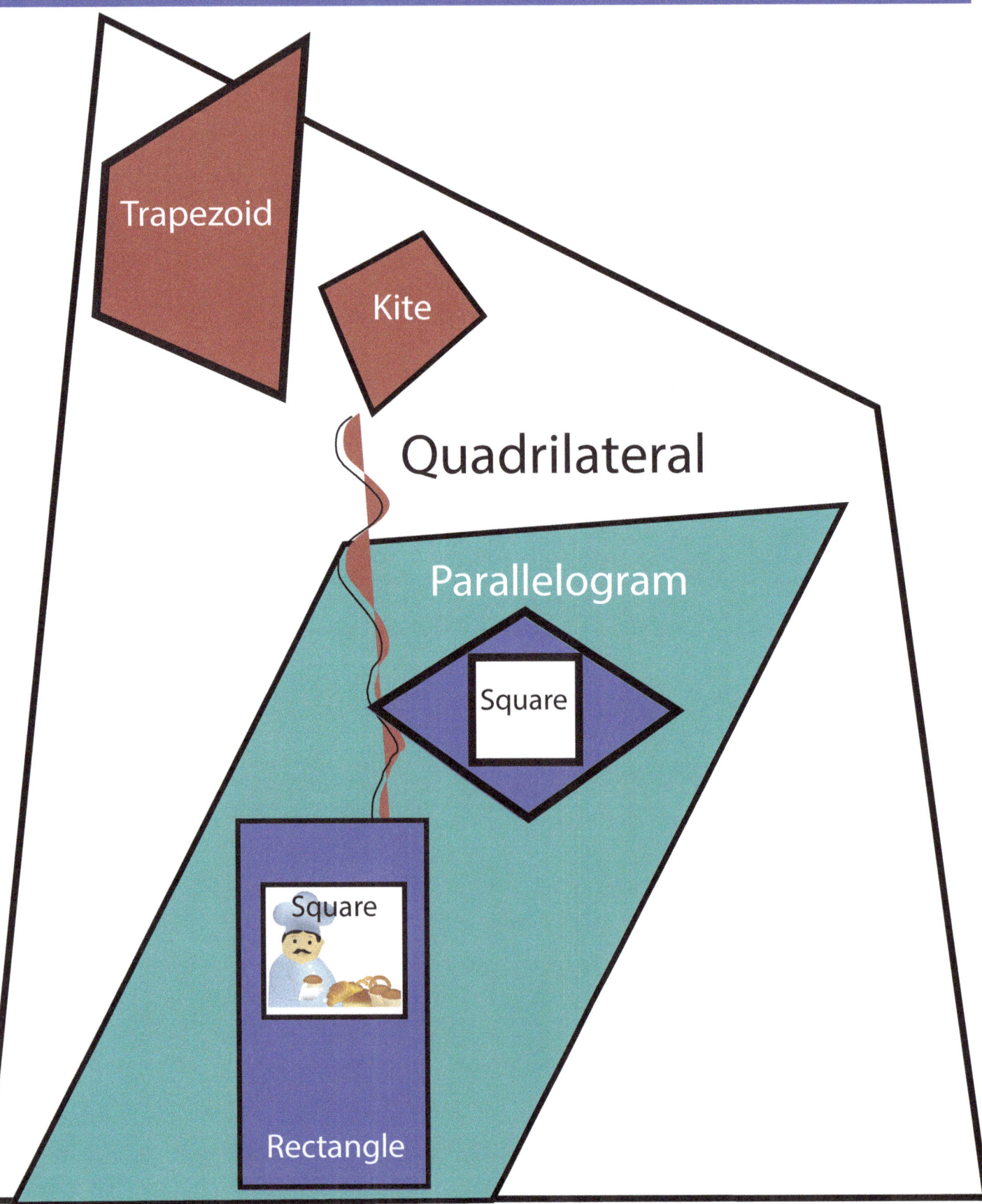

Do You Know the Quadrilaterals
Explore quadrilaterals in mathematics.

Oh do you know the quadrilateral, quadrilateral, quadrilateral?
Do you know the quadrilateral? A closed shape with 4 edges.
Oh, yes, I know the quadrilateral, quadrilateral, quadrilateral.
Oh, yes, I know the quadrilateral. 4 edges and 4 angles.

Now, do you know the trapezoid, the trapezoid, the trapezoid?
Do you know the trapezoid? It's a quadrilateral.
Oh, yes, I know the trapezoid, the trapezoid, the trapezoid.
Oh, yes I know the trapezoid. It has a pair of parallels. (a pair of LLs)

Now, do you know the parallelogram, the parallelogram, parallelogram?
Do you know the parallelogram? It's a quadrilateral.
Oh, yes, I know the parallelogram, parallelogram, parallelogram.
Oh, yes, I know the parallelogram. It has TWO pairs of parallels.

Now, do you know the rectangle, the rectangle, the rectangle?
Do you know the rectangle. It's a quadrilateral.
Oh, yes, I know the rectangle, the rectangle, the rectangle.
Oh, yes, I know the rectangle… (deep breath)
It has two sets of parallel edges where opposites edges and opposite angles are the same.

Now, do you know how to find a square, find a square, find a square?
Do you know how to find a square. It's a quadrilateral.
Oh, yes, we know how to find a square, find a square, find a square.
Oh, yes we know how to find a square. Look for all edges and all angles the same.

Now we know our quadrilaterals, quadrilaterals, quadrilaterals.
Now we know our quadrilaterals. Trapezoid, Parallelogram, Rectangle, Square!

ROOF ROOF
Explore pyramids.

Once upon a time there was Rhombi. She had a strong playhouse shaped like a cube. Autumn winds had not been able to push it down. Now that it was winter, she loved to hang out with a book and a warm blanket.

While the playhouse was snug and strong, the roof might need a redesign. Water had caused problems. Rain had made pools of water that sat on the roof. Snow and ice had frozen and thawed to make more puddles.

Drip. Drip. Drip. Freezing water and thawing ice had caused a hole in the roof! She worried that her books might be ruined. Rhombi imagined a waterfall in her house.

Rhombi's dad used a ladder to check out the playhouse roof. He looked over the flat roof full of puddles.

He said, "You have a real hole up there. I think you need to redesign the roof."

What kind of roof would she need? Rhombi stood outside with her umbrella and rain boots. She looked at her wet bird/squirrel feeder's steep roof. If she had a roof like that, water could not pool and make puddles.

She looked out beyond the pages of her world and asked, "Builders, can you help me build a waterproof roof for my playhouse?"

Water Is As Water Does
Explore the states of water in science.

Water Is As Water Does
Explore the states of water in science.

LIQUID
Water that we drink - or spill - is called liquid. Water changes names as it changes temperature.

ICE
Water that gets very cold turns solid. Solid water is called ice.

VAPOR
When water gets warm enough, it seems to disappear but is still in the air. Water we cannot see in the air is called vapor.

WATER, WATER EVERYWHERE
How many of these phrases do you know?

Wet your whistle...
Like oil and water...
Blood is thicker than water...
Mad as a wet hen...

CELSIUS
ice 0°
vapor 100°

FAHRENHEIT
ice 32°
vapor 212°

The Case of the Disappearing Sand
Explore the data collection in technology.

Read the charts with data collected by Rhombi and Pi. What story does the data tell?

What happened to the sand sculpture?

DAY 1 THE PYRAMID BASE IS 3 FEET SQUARE 4 FEET HIGH SUNNY WIND 5 MILES PER HOUR
DAY 2 BASE IS ALMOST 4 FEET SQUARE 3 FEET HIGH SUNNY WIND 10 MILES PER HOUR
DAY 3 BASE IS 2 FEET ON THE LONGEST SIDE BUT NOT ALL SIDES ARE EVEN. 2 FEET HIGH ON ONE SIDE ALMOST 3 FEET HIGH ON THE OTHER SIDE (NO LONGER SYMMETRICAL) RAIN LAST NIGHT WINDS 25 MILES PER HOUR
DAY 4 THE BASE IS NOT SQUARE 1 FOOT HIGH ON ONE SIDE ONE SIDE HAS LOST A LOT OF SAND RAIN WINDS GUSTING TO 4 MILES PER HOUR
DAY 5 MOSTLY FLAT AND NOT MUCH SAND LEFT

The Case of the Disappearing Sand
Explore data collection in technology.

The school bell rang. Kids tumbled out of the building. "What are you doing this weekend," Rhombi asked her friend Pi. Pi didn't have any plans. Neither did Rhombi. Maybe they would go for a swim or meet at the park? Rhombi and Pi saw a bright orange flier stuck to the school bulletin board.

"Sand Sculpture Con-test Saturday at 9AM"

The two friends looked at the sign. They looked at each other. They knew exactly what they'd be doing this weekend.

Rhombi and Pi hauled buckets, spades, small shovels, and lots of cardboard boxes to use as sand molds. The team got name tags and were assigned a 5ft by 5ft area of sand. They sketched out a quick design. Rhombi used clay to mold a model of the sculpture. When both friends were happy with the design, they started to build. They dug. They piled. They formed and pressed. Finally, the sand started to look like a pyramid. Small camels milled around at the base, and sand tents held tiny people.

About 4PM, Rhombi and Pi stepped back to look at their finished sculpture. They watched as judges made notes at each team's area. There were some amazing designs: seagulls flew over bridges; tsunamis crashed on beaches; castles held dragons, and a couple of robots held hands. The judges called all the contestants over to announce winners. Rhombi and Pi got bright green participation ribbons. They cheered as 1st, 2nd, and 3rd place winners given medals. Then everyone headed home, tired and sandy but proud.

Rhombi and Pi passed the beach each day on the way to school. They noticed something interesting. The sculptures were changing shape. Each day, their pyramid design changed just a bit. The two decided to make a timeline that showed changes every day for a week. Each day, they made a fast sketch of the sculpture and jotted down notes to keep a good record. They also tracked rain with a measuring cup system and wind with an anemometer. The weather detectives could solve this case. What was making sand disappear?

Circle, Triangle, Square
Explore rooftops in engineering.

What is the shape of your rooftop?

Circle, Triangle, Square
Explore rooftops in engineering.

Circle, Triangle, Square
How many rooftops touch the air?

Round like the top of a castle turret.
Round like the top of a tower.
Round like the clock that's hand strikes each hour.

Slanted like the ridge of a Swiss chalet.
Angled like the top of a tepee.
Pointed like a spire in the city.

Flat like a hospital's helicopter pad.
Flat like the tar on a high rise.
Flat like the planetarium, mirrors pointed at the sky.

Circle, Triangle, Square
How many rooftops touched the air?

I Love Homes
Explore houses in art.

I Love Homes
Explore houses in art.

I love homes.
Tall homes… Small homes…
Round homes and square
Homes in big buildings with hundreds of stairs.

I love homes:
Blue homes… Yellow homes…
Red homes and white
Homes with black shutters to close out the night.

I love homes.
Prairie homes… rabbit holes…
Towers and castles
Yurts in the grassland and tents in the desert.

I love homes.
Stilt homes… Ranch homes…
Igloos and trailers
Maybe I'll live on a boat like a sailor.

I love homes.
Cottage homes… Apartment homes…
Cabins or condos
Tiny houses or duplexes lined up in neat rows.

I love everyone's home.

What kind of home do you love?

The Camel and the Pyramid
Explore patterns in mathematics.

The Camel and the Pyramid
Explore patterns in mathematics.

There is a camel that lives in the far away sand of the Egyptian desert. His tan fur is stiff and rough. It has to keep out the tiny sand crystals as they crash through air during windstorms. He can close his nostrils and a third eyelid to keep out the terrible sharp sand. He has a hump that stores fat. The camel can live for days without food as his body uses the hump's fat to stay alive. Camels are tough.

The camel's caretaker is 14 years old. The young camel's guide wears a kafiyeh, a head wrap that does the same job as his camel's rough hide and closed nostrils. The scarf wound around the boy's head keeps sand and sun away from the boy's face. Wind blows lightly at 9 or 10 miles per hour through most days, but windstorms from the desert can be harsh. The wind may whip around the pyramids at 150 miles per hour. This kind of wind is dangerous to boys, camels and pyramids.

The mahout knows that the pyramids will last a long time, though. As he guides tourists from all over world, he tells the history of these pyramids. He tells the families that The Great Pyramid was made of horizontal rows of bricks. About 2,500,000 bricks were stacked like blocks. Each brick weighed around 2.5 tons. That means each brick weighed more than a car or a full-grown male giraffe. Tourists, happy with his stories, gave him tips called baksheesh. The boy earned a lot of baksheesh to help care for his camel and help feed his family.

On this day in March, the winds are blowing. There are no tourists to hear his tales or ride his camel. The young mahout and the 800 pound camel head home to strong stables. They need to escape the sand and wind. As they plod away toward Cairo, the mahout looks over his shoulder at The Great Pyramid. It would not blow away in the storms, even though little bits of its faces were eroded by the constant wind. Thousands of years would pass before the wind and water slowly brought the pyramids to the ground.

The camel lowered his third eyelid, closed his nostrils, and headed home toward fresh water and good food and a dry place away from the itching, scouring winds.

UNIT 3
PET FINDS A HOME

PET FINDS A HOME

Once upon a time there was Rhombi. She had a sturdy playhouse shaped like a cube. A pyramid roof protected her playhouse the from the sun, wind, and rain.

Rhombi took a walk. She saw all kinds of pets with their people. There were puppies. There was a bunny. She even saw a guy walking with a lizard on his shoulder.

After seeing all those animals, Rhombi was excited about getting a pet. She visited the animal shelter. Pet lived in Rhombi's playhouse for several months. Rhombi's mom was allergic to cats, so Pet could not live in their home.

Things were great while Pet was little. As she grew, the cat needed more room to run and play. On the first Monday in May, Rhombi found Pet in the sink. "What are you doing, Pet," she laughed and scratched the kitty under her chin.

On the second Tuesday in April, Rhombi found Pet in the cabinet. Pet was a little sulky. She did not want to play.

Rhombi started to worry about Pet.

On the third Wednesday in April, Rhombi hopped to stay on her feet when Pet scooted across the room. Pet grumbled a little. On the fourth Thursday in May, Rhombi tripped over Pet three times.

By the fourth week in April, Rhombi and Pet were grumpy about the tiny space. On Sunday of the same week, Rhombi made a decision. For Pet's sake, she needed to give the kitty more space to play.

Rhombi realized that she needed a special place for her new pet. She looked out beyond the pages of her world and asked, "Designers, can you help me build a small house for Pet?"

Meghan Finds a Purrl
Explore perspective in science.

One rainy day, Meghan and Davis were splashing around in mud puddles. They saw a kitten curled up in a cold, sad little ball by the curb. It looked miserable and wet. Meghan tried to pick it up, but the kitten hissed and tried to scratch her arm. Davis pulled off his sweatshirt. They wrapped the angry kitten in Davis's warm sweatshirt. Thank goodness the kitten calmed and began to purr. It fell asleep.

Meghan washed and brushed the little black ball of fur. She sat in front of the heater's vent with the kitten on her lap. It was so soft!

The veterinarian talked to the Davis and Meghan about healthy food and habits. He made them promise to have the kitten registered and vaccinated. That would keep the kitten from getting diseases later in life. The veterinarian even helped them pick kit-ten food. They picked a soft food for nutrients and a crunchy food to keep the kitten's teeth healthy.

The kids named their new friend Purrl and got her a bright orange collar. Meghan gave the kitten toys and a bed. She watched Purrl grow taller, and she noticed one day that her pet's eyes had turned green. Purrl got patch of white fur grew on her chest and white "socks" on her paws. Purrl was growing quickly.

Purrl slept on the Meghan's pillow every night and begged for breakfast scraps in the morning.

As Meghan held her purring cat, she whispered, "They all lived happily ever after. Good night, Purrl."

Purrl Finds a Home
Explore perspective in science.

Once upon a time, there was kitten. She was lonely and cold. Rain fell on her fur and dripped in her sad blue eyes. She huddled at the curb trying to get out of the wind.

She saw two giants looking down at her. Cold wet dripped down from giant shiny bodies. Their giant hands reached for her, and she hissed. The kitten's fur stood up in tufts, and her tail stuck straight out behind her body. She hopped, trying to look big and scary. The world got dark and a little bit warmer. Still, the kitten shook and hissed and meowed. She struggled to get free of the dark dry thing, but it was warm and dry. The kitten calmed and began to purr. She was dry. She was warm. She slept.

When she woke, the giants petted her fur. She ate some yummy brown paste and crunchy cookie food. She drank clear, cool water. She curled up by a hot wind and fell asleep again.

The kitten woke to find a thing around her neck. She clawed at the thing. It made jingly noises. She chased it in a circle. The giants laughed. She hissed, but she didn't really mean it…

The kitten washed her fur every day. She grew as tall as the giant's knees. The giants got names. The one with long head fur that smelled like vanilla was Meghan. The one with short head fur that smelled like grass was Davis. Grassy David visited a lot.

Purrl slept most of the day waiting in the window for sun to touch her paws. That's when the big stinky thing spit Meghan out onto the sidewalk every afternoon. Then she and her Meghan played until bedtime.

Purrl purred as the two best friends settled in for the night.

Rhombi Lived in a Zoo
Explore habitats and data collection in technology.

Rhombi Lived in a Zoo
Explore habitats and data collection in technology.

There once was a Rhombi who lived in a zoo. She had 26 rescues and life was a hoot. She fed them all dinner with fruits, veggies, protein, dairy, and bread. Rhombi checked on each one before going to bed.

Anna the ape was tucked in her nest.
Bella the boa liked sun warmed rocks best.
Calvin the colt needed plenty of hay.
Dobbie the deer slept someplace new every day.
Ella the Eastern Gorilla slept under the stars.
Fred the ferret curled up in her arms.
Gina Giraffe slept standing up tall.
Hannah the hedgehog curled up very small.
Iggy the inchworm had a leafy bower.
Jasper the jaguar was just waking up at this hour.
Kara the kitten had a basket and blankie.
Her friend Latrice Lemur climbed a tree with her hankie.
Mable the mastiff took up a whole bed.
Nonny the newt liked her fishbowl instead.
Odetta Otter burrowed into a holt.
Polly the Pig gets a wallow or away she will bolt.
Qbert the quail must sleep alone,
But Rory the rabbit wants friends in his home.
Selena the sloth bunks down on a tree limb.
Tito Terrapin carries his bed around with him.
Una Umbrella Bird has a blanket in her 48 inch cage.
Vixen the fox prefers a crevice with sage.
Wally the wren chooses a nest up high.
XRay the tetra sleeps surrounded by bubbles all the time.
Yuri the yak flops in the dirt on his back.
Zen Zebra curls up in the hay.

There once was a Rhombi who lived in a zoo. She had 26 rescues and life was a hoot. Rhombi makes notes in her file each day. She updates the spreadsheet to show eating habits and play. Who is not sleeping? Who bumped its head? Which animals might want shade or sun instead? Rhombi finished her work, shut down her computer and crawled into bed. "I'll do it all again tomorrow," she happily said.

It All Adds Up
Explore 3D printing in engineering.

Peg watched her dad add layer after layer to the towering party cake. Cake, icing, cake, icing, cake, icing and on and on until the cake was ready to serve. Dad carefully slid the cake off its platform. He removed the little stilts that held the cake still as he iced the layers. Finally, Dad and Poly delivered the finished dessert to the museum for a party.

Dad dropped Peg off at school on the way back to the bakery. Peg ran to her classroom. She was excited about designing for the cool little machines. When Peg arrived at her station in the classroom, she sat down at her wiggly table. She stuffed a piece of folded paper under one of the legs and tried not to get annoyed with the wobbling.

It All Adds Up

Explore 3D printing in engineering.

After fixing the table, Peg found a note. It read, "Innovators look for new ways to solve problems. How will you improve the classroom using our new technology? Sketch. Model. Design. Create!"

Peg's teacher introduced the new printer technology, and explained that these machines were not like a 2D paper and ink printer. 3D Printers used Computer Aided Design (CAD) files to make computer models. 3D printing turns the CAD computer models into real physical things. They melt mate-rials into thin layers on a surface, adding layer on layer until the full object is made. It sounded sort of like mak-ing a cake with Dad. He sketched and designed and layered. Leaning on her table, Peg's elbow slipped off as the table wobbled again. "Ugh, this table is so distracting," fumed Peg.

Suddenly, she had an idea. Peg got down on her stomach to take a closer look at the table leg. It was shorter than the other legs, and that made the whole table wobbly. Peg grabbed her logbook and started drawing. She remembered the little stilts her dad used to even out cakes for icing.

She measured the length of the other three legs, and drew a sort of stilt for the table. Peg got modeling dough to make a 3D model of her idea. She asked for help in using the scanner and watched as the CAD file was created to turn her idea into a program the printer could read. Her scanned clay model was digitally sliced into thousands of layers in the CAD program. Each of those layers would add up to make her stilt.

Peg held her breath as the first layer of her "Peg Leg" design was melted onto the 3D printer's surface. Slowly, layer by layer, the object was made. At the very end of the day, Peg pulled her "Peg Leg" from the printer's shelf. She broke off the little filaments that held it to the shelf, and hurried to her table. Kneeling, Peg slipped the 3D printed plastic piece under the short table leg. No more wobbly tables! As she headed out to meet her Dad at the car, Peg wondered how she could make her "Peg Leg" even bet-ter. Maybe there was a way to change the height? Maybe she could design different shapes for square or round legs. The ideas were endless, and they just kept adding up…

M.C. Escher's World
Explore tessellations in art.

"I don't grow up. In me is the small child of my early days" -M.C. Escher

M.C. Escher's World
Explore tessellations in art.

Escher was a well-known graphic artist. He composed visual material for printing. Escher lived from 1898 to 1972. Like DaVinci and Michelangelo, M.C. Escher was a leftie. He used his left-hand to write and do most activities. Also like these great artists, Escher made a lot of art!

In his lifetime, the graphic artist made 448 lithographs, woodcuts and wood engravings and over 2000 drawings and sketches. Escher also illustrated books, designed tapestries, created postage stamps, and painted murals.

After visiting a Moorish castle called Alhambra, Escher became interested in the tiled floors and ceilings. He started drawing special patterns called tiling or tes-sellations. Escher filled five notebooks with 137 of these tessellation drawings.

Tessellations (or tiling) are created when a shape is repeated over and over again. Each shape fits right up against the next to fill an entire surface. The shapes do not overlap or leave space between them. He described ways to make a tessellation pattern.

You can make pattern pictures like Escher's tessellations using polygons. When you've practiced filling an entire page with geometric shapes, you will be ready to try animals, plants, and trees.

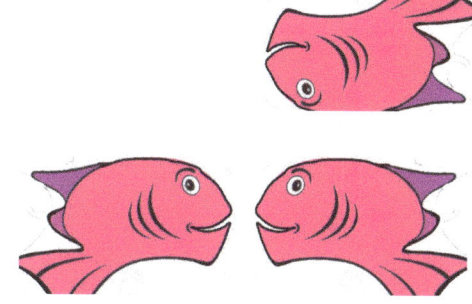

ROTATION
turn around a center point

REFLECTION (FLIP)
flip over a line

TRANSLATION
move without rotation or reflection

How Big Is a Guinea Pig

Explore size in mathematics.

How Big Is a Guinea Pig

Explore size in mathematics.

Food, Water, Air, Home: Guinea pigs never live alone. In groups of 5 or 10, wild guinea pigs use ½ inch claws to dig burrows in the ground for sleeping and keeping their tiny pups safe. Sometimes, the little communi-ty will use burrows left by other animals or bed down in rocky crevices. The guinea pigs work as a team, screaming to warn each other of predators or other danger.

Like rabbits, guinea pigs have incisor teeth that keep growing through their lives. Eating coarse grasses helps wear down the incisors. Premolars and molars crush and grind food. Also like rabbits, guinea pigs are actually rodents. They are not pigs at all!

Guinea pigs are mammals. Babies, called pups, drink mama sow's milk. Guinea pigs have fur, and they are born live and ready to wiggle. Pups grow quickly. From a few ounces to 2 or 3 pounds in less than a year, a full grown adult might be 10 to 14 inches long. That's about the length of lined paper or legal paper.

Guinea pigs live in their communities eating grass, sleeping through the day in safe burrows, and breathing the clear mountain air of South America from Columbia to Argentina.

Now that you know how guinea pig rodents live in the wild, how would you make a guinea pig happy in a classroom habitat?

HAPPY BIRTHDAY, PET

Explore habitats, polygons and structures.

Once upon a time there was Rhombi. Rhombi had a great playhouse. Pet had a house, too. Both buildings would not be toppled by Autumn winds. The pyramid tops let winter rain and snow slide right off the roof.

Pet had lived with Rhombi for many months. The cat would have her 1st birthday on April 15th. After going to the circus in early April, Pet and Rhombi chose a "Big Top" party theme.

Rhombi's was planning Pet's Summer birthday party and wanted to invite all of her barking, squawking, hissing, and mewing friends. Rhombi and Pet printed circus party invitations on primary colored paper.

Pet wanted 7 friends to attend her party. Rhombi invited a dog clown and ordered special cupcakes safe for pets! The third time Pet added a friend to the list, Rhombi realized that she didn't have enough room for a house full of 10 furry, scaly, feathery friends.

Each animal would need four square feet of space to be comfortable. She knew to ask for help. She looked beyond the pages of her world and asked, "Designers, will you help me design and build a covered space for Pet's birthday party?"

That's Bananas
Explore foods in science.

Bananas are a giant herb. They do not grow on trees.

A bunch of bananas is called a hand.

The word banana comes from an Arab word "banan" meaning finger.

Ecuador is the world's leading exporter of bananas.

An individual banana is called a finger.

That's Bananas
Explore foods in science.

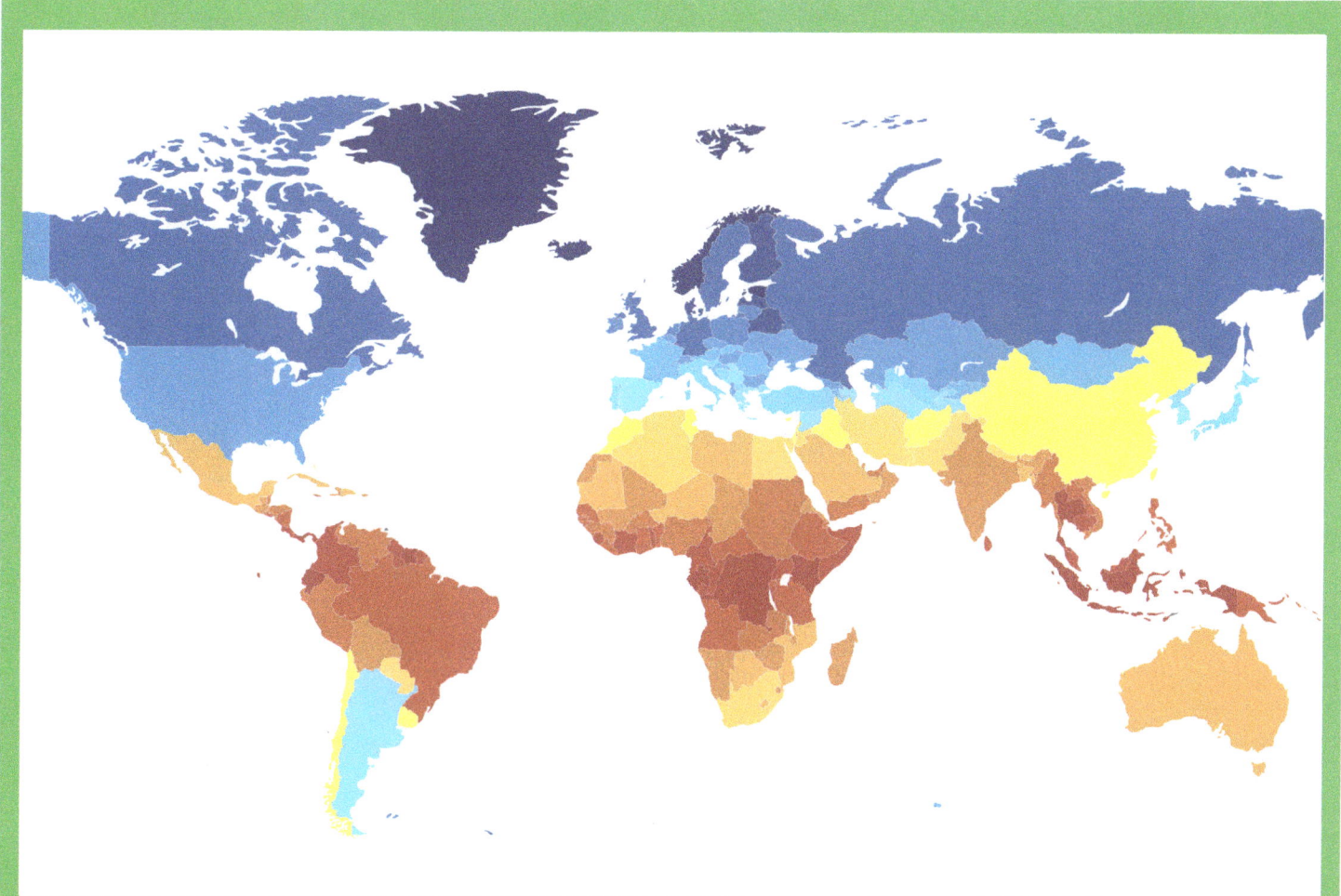

Most of the bananas you buy are grown within 20 degrees on either side of the Earth's equator in Costa Rica, Guatemala, Honduras, and Panama.

Waxing Colorful
Explore wax in technology.

If you wanted to build a house, what shapes would you use?

Waxing Colorful
Explore wax in technology.

In the hive (or in a wild nest), there are three types of bees:
1. a single female queen bee
2. up to 2000 male drone bees
3. some 20,000 to 40,000 female worker bees.

The worker bees raise larvae and collect the nectar that will become honey in the hive. When they leave the hive, they collect sugar-rich flower nectar and return.

Honey bees are a little like nurses, a little like architects, and a little like builders. They created wax to build honeycomb in regular hexagons. The honey bees use a shape that fits together as well as a puzzle. There are no holes or gaps between hexagons in the honeycomb. No wax-making honey or energy is wasted. That also makes honey bees sort of like engineers learning the best way to do a job.

The honeycomb "house" idea is working for humans, too. A company in the United Kingdom builds hexagonal houses. Each room of the house fits tightly together to save resources. These houses work for people the way a honeycomb works for bees. The honeycomb shapes take less energy to build, heat, and cool.

People have many uses for the beeswax that bees will not use. They make a lot more than any hive needs. Bee farmers harvest the rest for things like candles. Wax from honeybees is also used to make crayons. Most crayons are made of paraffin, but some organic companies are making crayons that contain fewer chemicals. Beeswax is mixed with pigments to make a rainbow of colors. You might be using beeswax candles or crayons in your house!

Strength in Numbers
Explore columns in engineering.

Visit http://alturl.com/aq5vy to make a Log Cabin.
Visit http://www.artistshelpingchildren.org for folded boats and lots of other great ideas.
Visit http://www.makingfriends.com to make hats in under a minute!
Visit http://www.creativekidsathome.com to make a dome.

Strength in Numbers
Explore columns in engineering.

Ginny kicked up dust with her sneakers as she headed down the dirt road. She was just wondering if the day could get any hotter when a motion caught her eye. Something was fluttering in the ditch to her left. She walked closer to get a look. She saw that it was most of a newspaper. Ginny did not want to leave litter in the road, so she rolled the paper into a tube and tucked it under her arm. She'd throw it away later.

She was about halfway to her friend's place when the sun broke free of the clouds. It was hot! It was sunny! The glare made her head ache. Ginny remembered the newspaper. She took a page and folded it into a paper hat. Her class had made them for a play last year. She never thought that skill would come in handy. Whew! Her face was cooler in the hat's shade.

Ginny pretended she was a pirate in her buccaneer's hat. She found a stick and heaved it around like a sword. Ginny was doing a pretty good pirate's "arrrr" when she crossed the little bridge over the Spring Road stream. Yesterday's rain had raised water level, and there was a great current. Ginny stopped to unfold two pages of the newspaper. She folded two small paper boats and set them gently down in the water on the North side of the bridge. She ran to the South side of the bridge to see which boat would arrive first.

She turned back to the road and finished her walk to Meghan's house. Just as she turned into her best friend's driveway, Ginny saw Meghan's father looking for something in his car. Ginny was curious.

"What are you looking for, Mr. Harris?" asked Ginny curiously. He showed her a letter that needed an envelope. Ginny had a handy solution. She used the last page of her newspaper to fold a quick envelope for Mr. Harris. He uses a sticker to close the flap, added a stamp and mailed his letter.

He asked, "Why are you carrying that old newspaper around?"

Pirate Ginny answered, "Because, it's not just a newspaper..."

Square in the Middle
Explore mosaic patterns in art.

Square in the Middle
Explore mosaic patterns in art.

Quad sat glumly on the bench. There was a box on the ground between his feet. When Rhombi sat down, she looked in the box.

"Quad, why are you hauling a box of broken pottery and dishes around the school?"

Quad looked into the box and put his head in his hands. His elbows rested on his knees, and he groaned. "That was my birthday present for Grandad. I dropped it on the way out of the art room, and I don't have enough money to buy another one."

Rhombi pulled the box closer and pushed some of the shards with her pencil. She closed her eyes and began to brainstorm. What could you do with a box and broken plates? Rhom-bi thought about broken things. Where had she seen broken plates in the last week?

"I have it," cried Rhombi loudly. She giggled and said, "Sorry, that was really loud, but I have an idea. "Quad's friend told him about the table her parents had bought for the patio. It had a pattern glued to the top in the shape of the moon and stars. The artist had glued each piece down and filled the spaces with grout. Tiling grout is usually based on cement and can come in color. You may see it between tiles in a kitchen, bathroom or floor.

"You could do the same thing with this box," said Rhombi. "Draw a design, and fill it in with the broken pieces."

Quad sat up and grinned. He liked the idea. Grandad loved cats. Quad could draw a cat on the box. He grabbed the box and headed toward the art room. Turning, he said to Rhombi, "You are a genius!" She smiled and waved him away.

Quad drew a cat. He glued all the broken pieces to the box. The art teacher gave Quad some grout to fill in the cracks between pieces. Mr. Kandinsky said that the grout would be dry in 24 hours. Quad left his masterpiece in the art room window. He was happy with the gift he'd give Grandad. He knew Grandad would love that he'd made the present all by himself.

Comin' Up a Cloud
Explore materials in mathematics.

A cotton boll weighs about 4 grams. A paper clip weighs about 1 gram, so a cotton boll is equal to about 4 paper clips in weight.

Comin' Up a Cloud
Explore materials in mathematics.

Ginny carefully wove through the rows of cotton. The Georgia weather was hot and humid. Ginny felt like her clothes were sticky, and the tiny gnats made mad little dashes toward her face and ears. She fanned them away. They didn't bite, but the tiny flies were annoying.

She'd kept an eye on the cotton since planting season in February. The flower buds had opened, and she'd watched as the petals fell to the dusty ground. With petals gone, the rest of the plant had ripened and grown. Each plant had a lot of cotton "bolls" with fluffy fibers.

This week, she and her family would hire local help to harvest the bolls. Then the piles of not-so-fluffy cotton would be "ginned" to get rid of all the seeds. They would send the cotton off to a mill. The mills will mix and clean the cotton. Picking machines will break the piles of cotton into pieces. Dirt has to be shaken out of the cotton fibers.

A carding machine that separates the fibers will comb through the piles. Combing makes sure all the fibers run in the same direction. Ginny's Great-Great Grandma used to do all this by hand, but the machines work with a lot more fiber in a much shorter amount of time.

Great-Great Grandma spent weeks doing what a machine finishes in an hour. A long, smooth rope called a "sliver" forms. The "sliver" has to be pulled and twisted until it's thin. The thin "roving" is finally ready for a spinning frame. Before it becomes string, all that beautiful roving is twisted and wound onto bobbins.

Ginny wondered where the cotton boil she touched would go. Would it end up on a cargo ship bound for China? Would it land in a pretty gift store? Would she wear it next year as a t-shirt or a pair of jeans? Maybe it would wind her yo-yo or fly a kite.

As Ginny bent down to tie her cotton shoelace, a cloud crossed the field. She glanced at the sky. It looked like it was "comin' up a cloud."

Read more : http//www.ehow.com/about_6360791_cotton-made-thread_.html

EXTENSION
RHOMBI'S CANDIES

Rhombi's Candies
Module 2: Extension Unit

Once upon a time, there was Rhombi. Rhombi had a wonderful friend named Pet and a cool cube playhouse with a strong pyramid roof. Pet even had her own house and a party pavilion for friends!

Pet and Rhombi loved 2 dimensional and 3 dimensional shapes of all kinds. Their house was filled with interesting squares, triangles, hexagons, cubes, and pyramids. Rhombi's favorite decoration was a cubist painting in a trapezoid frame.

Rhombi's friend Pet had celebrated her birthday just a few months ago. They had served circus food including cotton candy, lollipops, popcorn, and other pet friendly candy. It had been so popular that Rhombi wanted to recreate the menu at the market.

Rhombi's parents had given permission for Rhombi to open a candy store cart. The cart could be wheeled down to the Saturday farmer's market. When the market ended, the cart could be wheeled right back home.

The farmer's market hosted lots of different foods and crafts. Rhombi's friends Pi and Radius sold fruits and vegetables at the market. She had asked to borrow the cart one weekend and try her business idea.

Pi and Radius had to visit Grandma, so they loaned Rhombi their cart one weekend. She had piled it high with candy and cookies. The cart was open, and all the chocolate melted. Rhombi decided that she needed a new design for her candy cart.

Rhombi looked beyond the pages of her world and asked, "Builders, will you help me design and build a candy store cart with the materials we have?"

KINDERGARTEN
VOCABULARY

DATA	SCIENCE INFORMATION
CONCLUSION	THE LAST PART
PREDICT	TO SAY WHAT YOU THINK WILL HAPPEN
DESCRIBE	SAYING OR DRAWING WHAT YOU SEE, HEAR, TOUCH, TASTE OR SMELL
OBSERVE	TO SEE, HEAR, TOUCH, TASTE OR SMELL
INVESTIGATE	TO GATHER INFORMATION
RECYCLE	TO USE AGAIN, ESPECIALLY TO REPROCESS
DISPOSE	TO GET RID OF AN OBJECT
REUSE	TO USE AGAIN, ESPECIALLY AFTER SALVAGING OR SPECIAL TREATMENT OR PROCESSING
GOGGLES	A PAIR OF TIGHT FITTING EYEGLASSES, OFTEN TINTED OR HAVING SIDE SHIELDS, WORN TO PROTECT THE EYES FROM HAZARDS SUCH AS WIND, GLARE, WATER

PHYSICAL PROPERTIES

APPEARANCE	THE OUTWARD LOOK OF AN OBJECT: COLOR, SHAPE, SIZE, ETC.
FIVE SENSES	TASTE, SMELL, SEE, HEAR, TOUCH
HAND LENSES	A HAND HELD MAGNIFYING GLASS
MASS	AMOUNT OF MATTER IN OBJECT OR SUBSTANCE, MEASURED IN GRAMS
MATTER	ANYTHING THAT HAS MASS AND TAKES UP SPACE
NOTEBOOK	A BOOK OF BLANK PAGES FOR NOTES
OBJECT	THING
PATTERN	SHAPES, COLORS, OR LINES PUT TOGETHER IN A CERTAIN WAY
PHYSICAL PROPERTIES	ATTRIBUTES THAT CAN BE OBSERVED AND MEASURED

TOOL	SOMETHING USEFUL
COOLING	TO MAKE LESS WARM
ENERGY	THE ABILITY TO CAUSE MOVEMENT OR CREATE CHANGE
EVAPORATION	THE CHANGE FROM A LIQUID TO A GAS
FREEZING	TO CHANGE FROM A LIQUID TO A SOLID BY LOSING HEAT
HEATING	TO MAKE WARM OR HOT
MELTING	TO CHANGE FROM A SOLID TO A LIQUID BY ADDING HEAT
TEMPERATURE	HOW HOT OR COLD SOMETHING IS
THERMOMETER	A TOOL USED TO MEASURE TEMPERATURE
PATTERN	EVENTS THAT REPEAT

DAY	THE TIME BETWEEN SUNRISE TO SUNSET
NIGHT	THE TIME BETWEEN SUNSET TO SUNRISE
CHANGE	TO GIVE A DIFFERENT FORM OR APPEARANCE TO

KINDERGARTEN
VOCABULARY

MOTION	
MOTION	A CHANGE IN POSITION
DIRECTION	THE PATH WAY OF AN OBJECT IN MOTION
ABOVE	ON TOP OF; HIGHER THAN
BELOW	UNDER; UNDERNEATH
BEHIND	AT THE BACK OF; LAST
BESIDE	NEXT TO
STRAIGHT	NOT BENT OR CURVY
ZIGZAG	MOVING WITH SHARP TURNS
BACK AND FORTH	MOVING SIDE TO SIDE
ROUND AND ROUND	MOVING IN A CIRCLE

MATTER	ANYTHING THAT HAS MASS AND TAKES UP SPACE
PHYSICAL PROPERTY	ATTRIBUTES THAT CAN BE OBSERVED AND MEASURED
SHAPE	THE OUTWARD FORM OF AN OBJECT
MASS	THE AMOUNT OF MATTER IN AN OBJECT
TEMPERATURE	A MEASURE OF THE AMOUNT OF HEAT
TEXTURE	HOW A SUBSTANCE FEELS
ENERGY	THE ABILITY TO DO WORK
FIVE SENSES	TASTE, SMELL, SIGHT, HEAR, TOUCH
LIGHT ENERGY	ENERGY THAT YOU CAN SEE

ASTRONOMER	SCIENTIST THAT STUDIES CELESTIAL PHENOMENA
CONTRACTOR	A GENERAL CONTRACTOR IS RESPONSIBLE FOR THE DAY TO DAY OPERATIONS OF A CONSTRUCTION SITE. CONTRACTORS MANAGE SALES PEOPLE, BUILDERS, PLUMBERS, ELECTRICIANS, AND ALL THE OTHER TRADES REQUIRED TO COMPLETE A PROJECT. THE CONTRACTOR ALSO COMMUNICATES INFORMATION TO ALL THE PEOPLE INVOLVED IN THE PROJECT.
DESIGNER	INDUSTRIAL DESIGNERS DEVELOP CONCEPTS AND SPECIFICATIONS THROUGH COLLECTION, ANALYSIS AND SYNTHESIS OF DATA GUIDED BY THE SPECIAL REQUIREMENTS OF THE CLIENT OR MANUFACTURER. DESIGNERS MAKE CLEAR AND CONCISE RECOMMENDATIONS THROUGH DRAWINGS, VERBAL DESCRIPTIONS, AND MATH MODELS.
ENGINEER	PROFESSIONAL THAT APPLIES PRINCIPLES OF SCIENCE AND MATHEMATICS BY WHICH THE PROPERTIES OF MATTER AND THE SOURCES OF ENERGY IN NATURE ARE MADE USEFUL TO PEOPLE.